$26

Open a Newspaper

by Susan Korman

BLACKBIRCH®
PRESS

THOMSON
★
GALE

San Diego • Detroit • New York • San Francisco • Cleveland • New Haven, Conn. • Waterville, Maine • London • Munich

For more information, contact
The Gale Group, Inc.
27500 Drake Rd.
Farmington Hills, MI 48331-3535
Or you can visit our Internet site at http://www.gale.com

Photo Credits: see page 47.

LIBRARY OF CONGRESS CATALOGING-IN-PUBLICATION DATA

Korman, Susan.
 Open a newspaper / Susan Korman.
  p. cm. — (Step back science series)
 ISBN 1-56711-677-9 (hardback : alk. paper)
 [1. Papermaking.] I. Title. II. Series.

TS1105 K677 2003
676'.286—dc21   2002013546

Printed in United States
10 9 8 7 6 5 4 3 2 1

Contents Open a Newspaper

How to Use This Book

Each Step Back Science book traces the path of a science-based act backwards, from its result to its beginning.

Each double-page spread like the ones below explains one step in the process.

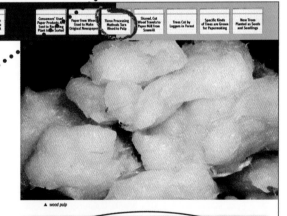

A timeline along the top describes all the steps in the process. A marker indicates where each spread is in the process.

A question ends each spread and is incorporated into the title of the next spread.

A short description gives a quick answer to the question asked at the end of the previous step.

Sidebars show interesting related information.

Side Step spreads, like the one below, offer separate but related information.

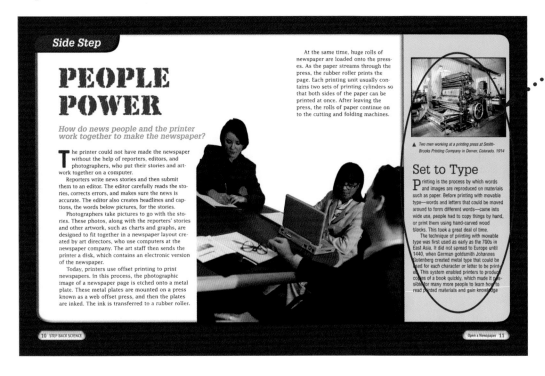

The Big Picture, on pages 40–41, shows the entire process at a glance.

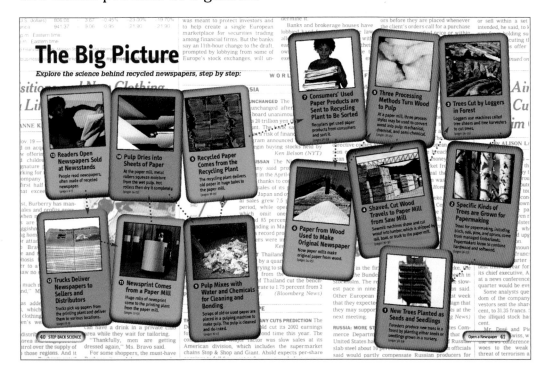

| Readers Open Newspapers Sold at Newsstands | Trucks Deliver Newspapers to Sellers and Distributors | Newsprint Comes from a Paper Mill | Readers Open Newspapers Sold at Newsstands | Pulp Mixes with Water and Chemicals for Cleaning and Bonding | Recycled Paper Comes from the Recycling Plant |

Readers Open Newspapers Sold at Newsstands

How are newspapers produced and brought to newsstands?

Chances are, you see a newspaper every day. In the United States alone, millions of newspapers are printed each day, and that takes tons of paper. That is why newspapers today are usually printed on recycled, or used, paper.

Before this recycled newspaper arrives on newsstands, it makes a long journey from a reader's hands to a recycling plant to a paper mill and finally to the printer.

Yet the paper's history actually stretches back to its humble beginnings on Earth as a tiny seedling ready to grow into a tree. Where the paper has been is as good a story as any article inside.

How do recycled newspapers get to readers?

Consumers' Used Paper Products Are Sent to Recycling Plant to Be Sorted

Paper from Wood Used to Make Original Newspaper

Three Processing Methods Turn Wood to Pulp

Shaved, Cut Wood Travels to Paper Mill from Sawmill

Trees Cut by Loggers in Forest

Specific Kinds of Trees are Grown for Papermaking

New Trees Planted as Seeds and Seedlings

Trucks Deliver Newspapers to Sellers and Distributors

How do recycled newspapers get to readers?

Newspapers are delivered in bundles by trucks to newsstands, stores, vending machines, and other distributors. After newspapers have been printed at a printing plant, machines there cut and fold the paper into its individual sections. Finished newspapers are bundled together and stacked at the plant where drivers pick them up to deliver them to the places where they will be sold. The other distributors deliver subscription copies of the papers to homes, schools, and office buildings.

But where does the printer get the recycled paper to make the newspaper?

▲ *Many egg cartons are made of recycled paper.*

New Life for Old Newspapers

About 68 percent of all the newsprint used in the United States is made from recovered, or used, paper. About one-third of this newsprint is recycled back into newsprint. Other products made from recovered newsprint include egg cartons, printing and writing paper, cereal boxes, books, corrugated boxes, and insulating materials. Shredded newspaper is also used for animal bedding in places such as pet stores, animal clinics, and farms.

PEOPLE POWER

How do news people and the printer work together to make the newspaper?

The printer could not have made the newspaper without the help of reporters, editors, and photographers, who put their stories and artwork together on a computer.

Reporters write news stories and then submit them to an editor. The editor carefully reads the stories, corrects errors, and makes sure the news is accurate. The editor also creates headlines and captions—the words below pictures—for the stories.

Photographers take pictures to go with the stories. These photos, along with the reporters' stories and other artwork, such as charts and graphs, are designed to fit together in a newspaper layout created by art directors, who use computers at the newspaper company. The art staff then sends the printer a disk, which contains an electronic version of the newspaper.

Today, printers use offset printing to print newspapers. In this process, the photographic image of a newspaper page is etched onto a metal plate. These metal plates are mounted on a press known as a web offset press, and then the plates are inked. The ink is transferred to a rubber roller.

At the same time, huge rolls of newspaper are loaded onto the presses. As the paper streams through the press, the rubber roller prints the page. Each printing unit usually contains two sets of printing cylinders so that both sides of the paper can be printed at once. After leaving the press, the rolls of paper continue on to the cutting and folding machines.

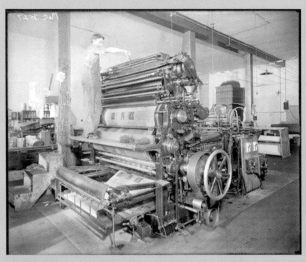

▲ *Two men working at a printing press at Smith-Brooks Printing Company in Denver, Colorado, 1914*

Set to Type

Printing is the process by which words and images are reproduced on materials such as paper. Before printing with movable type—words and letters that could be moved around to form different words—came into wide use, people had to copy things by hand, or print them using hand-carved wood blocks. This took a great deal of time.

The technique of printing with movable type was first used as early as the 700s in East Asia. It did not spread to Europe until 1440, when German goldsmith Johannes Gutenberg created metal type that could be used for each character or letter to be printed. This system enabled printers to produce copies of a book quickly, which made it possible for many more people to learn how to read printed materials and gain knowledge.

Newsprint Comes from a Paper Mill

Where does the printer get the recycled paper to make the newspaper?

The printer gets the paper from a paper mill. At the mill, when the paper first comes off the paper-making machine, it is wound around reels at a very high speed. This creates huge rolls of paper approximately 30 feet wide and with a weight of up to 20 tons. The rolls are then cut into smaller rolls so they can be easily handled and printed at the plant. Trucks deliver the smaller rolls to printing plants and other consumers.

So how does the papermaking machine create large sheets of recycled paper?

Huge rolls of paper wait to be cut into more manageable sizes. ▶

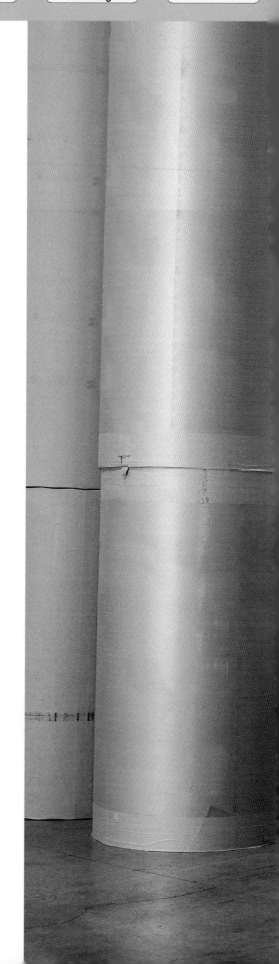

| Consumers' Used Paper Products Are Sent to Recycling Plant to Be Sorted | Paper from Wood Used to Make Original Newspaper | Three Processing Methods Turn Wood to Pulp | Shaved, Cut Wood Travels to Paper Mill from Sawmill | Trees Cut by Loggers in Forest | Specific Kinds of Trees are Grown for Papermaking | New Trees Planted as Seeds and Seedlings |

Cut to the Source

One of the simplest ways to cut paper is with scissors, but there are many different accounts about how the handy tool was invented. While some people credit the Italian artist Leonardo da Vinci with inventing scissors for the purpose of cutting art canvases, most historians agree that the earliest form of scissors dates back to 300 B.C. in ancient Egypt.

These early scissors were a simple version of what is used today. Instead of having two crossed blades connected by a screw, these scissors were made of one piece of metal with two blades connected by a curved handle. The user had to press the blades together to make them work.

Evidence suggests that cross-blade scissors first appeared around A.D. 100 in Rome. The first written account of cross-blade scissors' use is from A.D. 500, when the writer Isidore of Seville described them as a tool for barbers and tailors.

Pulp Dries into Sheets of Paper

How does the papermaking machine create recycled paper?

At the mill, recycled scraps of paper are combined and flattened in a machine to make the large rolls of paper. Workers pour a wet, slushy mixture called pulp—made from recycled paper scraps, water, and chemicals—onto a metal box in the paper machine. From there, machines spray the mixture evenly onto a wide, flat, wire screen. The screen moves quickly through the paper machine, and water begins to drain from the pulp. The fibers left behind bond together on the screen to form a large, but still wet, sheet.

Several felt-covered rollers spin across the sheets, which travel quickly around the machine, and press out even more water. The wet sheets finally cross over a series of hot metal rollers, which completely dry them.

Why was the paper wet in the first place?

Hot metal rollers complete the process of turning wet pulp into dry paper. ▶

| Consumers' Used Paper Products Are Sent to Recycling Plant to Be Sorted | Paper from Wood Used to Make Original Newspaper | Three Processing Methods Turn Wood to Pulp | Shaved, Cut Wood Travels to Paper Mill from Sawmill | Trees Cut by Loggers in Forest | Specific Kinds of Trees are Grown for Papermaking | New Trees Planted as Seeds and Seedlings |

The Goods on Paper

It is obvious that some household items, such as printer paper and paper towels, are made of recycled paper. You may be surprised to learn that some of the objects below are also made from recycled paper or from the natural byproducts of papermaking.

Sponges	Roadside flares	Liquid soap
Emery boards	Sandpaper	Tea bags
Suntan lotion	Flashlight batteries	Coffee filters
Wicker furniture	Jigsaw puzzles	Postage stamps
Window shades	Crayons	Report cards
Sports helmets	Hairspray	Shaving cream
Luggage	Masking tape	Toothpaste
Lipstick	Kites	
Car wax	Stickers	

Pulp Mixes with Water and Chemicals for Cleaning and Bonding

Why was the paper wet in the first place?

Water and chemicals were added to the recycled scraps not only to clean them and make them stick together, but also to de-ink them.

The de-inking process removes printing ink and "stickies," sticky substances such as glue residue and adhesives, which may be left over from the paper's last use. Paper mills often combine two processes of de-inking to clean the pulp fully. Tiny bits of ink are removed in a rinse process called washing. Larger particles, including stickies, are removed by a process called flotation de-inking. During flotation de-inking, workers feed pulp into a flotation cell, a vat in which soap-like chemicals called surfactants are injected into the pulp. These chemicals loosen the ink and stickies, which then attach to the air bubbles that float to the surface. The air bubbles produce a foam that can be skimmed from the surface of the mixture. Clean pulp is left behind.

The clean pulp is refined, or beaten, so that the recycled fibers swell. The refining process also separates any large bundles of fibers that are still clumped together. Color-stripping chemicals are then added to the pulp to remove any dye. When white recycled paper is being produced, the pulp is bleached to brighten it. If brown recycled paper is being made, the bleaching process is not necessary.

So where does the paper mill get the used paper?

▲ *Recycled paper is mixed with hot water and cleansers in a container called a pulper.*

▲ *Cardboard boxes are recycled following the same process as old newspapers.*

Heavy-Duty Paper

Cardboard, also known as paperboard, is thick and strong. Because it is also lightweight, it is used for making boxes and other types of packaging.

Like other forms of paper, cardboard can be recycled. At recycling plants, cardboard is collected into huge bales and transported to a paper mill. There, the cardboard follows the same process as old newspapers and other recycled papers. It is placed into a pulping machine, and water and chemicals are added to make pulp. Next, the pulp is cleaned, de-inked, and dried. The dried cardboard is rolled onto huge rolls and then cut into smaller rolls for easier handling. The paper is sent to an outside factory to be layered and folded.

Recycled Paper Comes from the Recycling Plant

Where does the paper mill get the used paper?

The previously used paper comes to the mill from a recycling plant. When the paper first arrives at the mill, it is in huge bales that have been tightly compacted and bound with wire. Separate bales may consist of paper recovered from newspapers, magazines, and food and drink cartons, as well as individual sheets of computer and writing paper. The bales are very large and heavy; just one can weigh as much as a ton.

Where did the recycling center get the paper?

Various kinds of previously used paper arrive at the mill in large bales. ▶

Test Your Wood Fiber Knowledge

How many times can paper (in other words, its wood fibers) be recycled?

a. an indefinite number

b. 15-20

c. 5-7

d. 150-200

(See answer on page 47.)

Consumers' Used Paper Products are Sent to Recycling Plant to be Sorted

Where did the recycling center get the paper?

The recyclers pick up paper products from consumers— people and businesses that have used the products. In some parts of the country, people bring it to the center themselves. In other communities, the used paper products may be collected at homes and offices by trucks from the center, a trash collection agency, or a paper-stock dealer.

At the plant, used paper products are sorted into different categories. For example, used cardboard and corrugated packaging from supermarkets are separated from old newspapers and magazines. Paper that has been badly contaminated by food or is stuck to metal or plastic cannot be recycled, so it is discarded.

After the paper products are sorted, they are further separated into grades. Paper products of a particular grade have a similar texture and weight. The grade determines which bale the paper products will go into for the trip to the mill and what will be done with them there. In general, newspapers are made from other newspapers and magazines, but sometimes new paper is added to the mix.

Where does the new paper come from?

Families can help the recycling process by ▶
sorting used paper products into categories.

Consumers' Used Paper Products are Sent to Recycling Plant to be Sorted	Paper from Wood Used to Make Original Newspaper	Three Processing Methods Turn Wood to Pulp	Shaved, Cut Wood Travels to Paper Mill from Sawmill	Trees Cut by Loggers in Forest	Specific Kinds of Trees are Grown for Papermaking	New Trees Planted as Seeds and Seedlings

Save That Cereal Box!

At home, many things can be recycled. Recyclable paper items include newspapers, magazines, paper from computer printers, old envelopes, and junk mail. Items made of cardboard, such as cereal boxes, egg cartons, and other types of cartons can also be recycled.

Containers made of plastic, glass, and aluminum are also recyclable. They must, however, be rinsed and sorted according to what they are made of.

People can also recycle things at home or at school by finding new uses for them. For example, old yogurt containers are handy for planting seeds to grow herbs and other houseplants.

TALES OF TWO PAPER TYPES

Has all recycled paper been used by people or businesses?

Some recycled paper has never been used. Known as pre-consumer recovered paper, this type of recycled paper consists of trimmings and scraps from printing, carton manufacturing, and other converting processes. These scraps are reprocessed without ever being used by the public.

Post-consumer paper is paper that has been used and comes to the plant for recycling. It consists of things such as newspapers, magazines, used cartons, and corrugated boxes. While some paper is made of completely recycled paper, papermakers usually combine recycled fibers with new fibers to create papers of various grades of quality. Mixing the old fibers with new fibers also lengthens the life of the fibers, which can weaken over time.

Paper made from recycled fibers used to cost more than paper made completely from new fibers. Today, many recycled papers cost the same or less than new paper.

pre-consumer paper

post-consumer paper

Government Regulations

In recent years, environmentalists have pressured the lumber industry and governments to conserve forests and landfill space. This has led to many practices by the lumber industry and the U.S. government. One of these practices is known as forest certification, a system of identifying wood products that are manufactured according to certain environmental standards. Trade groups—organizations affiliated with particular industries—certify forest products for the paper industry. One of the most prominent of these certifying groups is the American Forest and Paper Association. Another influential certifying group is the Forest Stewardship Council, which certifies products themselves and oversees other certifiers.

The U.S. government has tried to encourage conservation by creating paper-use guidelines for its employees. Federal agencies are now required to use paper with a minimum of 30 percent post-consumer content for most uncoated printing and writing papers, and 10 percent post-consumer content for most coated papers. Many state governments and businesses have adopted this policy as well.

Paper from Wood Used to Make Original Newspaper

Where does the new paper come from?

At a new paper mill, which is often (but not always) a different kind of mill from the ones that process recycled paper, newly created paper is made from wood. Before the wood gets to the mill, it must be small enough to transport easily and fit through the mill machines. Although some companies use separate new paper mills for pulping and papermaking, larger mills handle both tasks. At these large mills, the wood arrives in the form of small logs or wood chips and is made into pulp.

How is wood turned into pulp to make new paper?

A machine dumps wood chips to be turned into pulp. ▶

Make Your Own Paper

You can make paper at home. Be sure to ask an adult to help.

Materials: Scrap paper torn in 1" x 1" pieces, Old clothing (to wear during the process), Disposable aluminum pan, 11" x 13", Wire mesh screen, Large tub, at least 2 gallon size, Dish towels, Blender, Sponge (for cleaning messes), Rolling pin, Iron, Strainer, Old newspapers.
Optional: short strands of thread or bits of dried flowers or spices

1. Making paper can get messy, so choose a suitable workspace.

2. Soak the scraps of paper overnight in the large tub filled with warm water.

3. Make a square cutout in the bottom of the aluminum pan. Leave about one inch along the outer edge of the pan. Cut a piece of wire screen large enough to fit over the cutout when placed in the bottom of the pan. (You have just created a deckle frame, which can also be purchased at crafts stores.)

4. Fill the blender halfway with warm water and add the wet scrap paper. Put the lid on the blender and blend at medium speed until there are no loose pieces of paper. (You can blend in the dried flowers or spices at this point.) Pour the blended mixture into the large tub and fill with warm water. Mix thoroughly until all the ingredients are dispersed.

5. Place the deckle frame into the tub. Gently move it back and forth to get an even layer of fibers on the screen. Then lift the frame from the mixture, keeping it flat. Allow the pulp to drip over the tub until most of the water has drained through the screen. Press the pulp gently with your fingers to press out excess moisture.

6. Place newspaper on a flat surface and flip the screen, paper side down, on the newspaper. Lift the screen gently, leaving behind the paper. Cover the paper with the dish towels, and then use the rolling pin to squeeze out the remaining water. Place the wet newspaper on fresh paper, and let the sheets dry overnight, leaving the dish towels in place.

7. When the paper is mostly dry, use an iron at a medium setting and iron the paper gently to completely dry it. Then pull the dishtowel gently from both ends, and stretch it to loosen the paper from the cloth. Peel off the paper carefully. You have made a sheet of paper!

Note: Any remaining pulp can be stored in a plastic bag in the freezer for future use.

A TRUE ORIGINAL

Is new paper made the same way as recycled paper?

The process for making new paper is very similar to that for making recycled paper. The pulp used to make new paper, however, comes from wood chips rather than wet scraps of used paper. In new paper processing, the first step is the harvesting of trees. The logs are then brought to a paper mill where they are cleaned and ground into chips. These small chips are then converted to pulp.

During this stage, the individual wood fibers in the chips are separated from one another. The pulp that results is wet and mushy. Next, the water must be squeezed from the pulp. As with recycled paper pulp, this is done with screens for draining and rollers for further pressing of the pulp. To finish the job, hot, steam-filled cylinders heat the paper and dry it thoroughly.

Proud Paper Producers

About 350 million tons of paper and cardboard are made each year throughout the world. The United States produces the most paper; U.S. mills churn out about 30 percent of the world's paper and cardboard.

Within the United States, Wisconsin is the biggest paper producer. More than 5.3 million tons of paper and 1.1 million tons of cardboard are produced there annually.

◀ From left to right, stages in the new paper process: harvesting trees, transporting the logs, grinding logs into chips, converting chips into pulp.

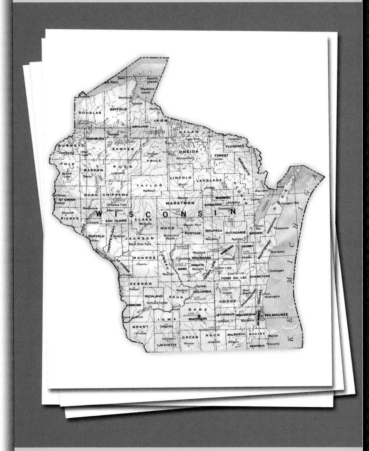

Three Processing Methods Turn Wood to Pulp

How is wood turned into pulp to make new paper?

At the new paper mill, wood turns into pulp by one of three processing methods: mechanical, chemical, or semi-chemical. Mechanical processes are the oldest way to turn cut wood to crumbs. In the stone ground-wood process, developed around 1840 in Germany, heat softens tree wood's lignin, a substance that hardens the cell walls of wood and strengthens it overall. The wood is then thrust repeatedly against a flat, circular stone called a grindstone, which revolves on an axle. The grindstone shreds loose fibers.

Another mechanical process is called thermo-mechanical pulping. In this case, a machine chops the wood into chips that are then heated and inserted between the rapidly spinning disk-like blades of a refiner machine. The disks break down the wood into fibers.

In chemical pulping, chemicals dissolve the lignin. The sulfite process is a popular chemical process in which wood is cleaned and cut into tiny chips before being heated in an acid solution that breaks them down further. In another chemical method, the Kraft (or sulfate) process, wood chips are heated and broken down in a solution of caustic soda and sodium sulfide.

Wood broken down in a semi-chemical process is commonly soaked in chemicals that soften the lignin and then chopped in a refiner. No matter what process is used to make pulp, the fibers come from pieces of wood, which are fairly small by the time they get to the mill.

But where is wood chopped down to size for travel to the paper mill?

▲ *wood pulp*

Ancient Papermaking

Papyrus, a tall plant grown in the Nile valley, was extremely important to ancient civilizations because they could make paper from it. In fact, the word paper derives from papyrus. To make a sheet of papyrus that was suitable for writing, Egyptians cut papyrus stalks into thin strips. Then they arranged the strips into criss-crossed layers, which they pounded and pressed into sheets. The flat sheets were polished smooth by stones, shells, or bones.

▲ *a sheet of papyrus*

| Readers Open Newspapers Sold at Newsstands | Trucks Deliver Newspapers to Sellers and Distributors | Newsprint Comes from a Paper Mill | Pulp Dries into Sheets of Paper | Pulp Mixes with Water and Chemicals for Cleaning and Bonding | Recycled Paper Comes from the Recycling Plant |

Shaved, Cut Wood Travels to Paper Mill from Sawmill

Where is wood chopped down to size for travel to the paper mill?

Logs and wood pieces are cut into smaller pieces at the sawmill. There, one machine removes tree bark while another electronically scans the shaved logs and wood pieces to calculate the easiest way to cut them. Some wood is ground into chips or small logs, but some paper mills may get wood in the form of boards that can fit easily through paper mill machines.

To create the boards, rows of saws, called gang saws, slice the logs. Then a conveyor belt carries the boards to a set of smaller saws that squares the ends of the boards and trims them into specified lengths. The boards, now known as rough lumber, move along on another belt where workers inspect them for size and quality. At this point, inspectors decide if the lumber is sturdy enough and suitable to be used as building material or if it is soft enough to be made into paper products. After it passes inspection, the lumber is ready for the paper mill.

Lumber that is shipped long-distance to paper mills travels by rail. Boats and trucks, however, are used when distance or inaccessibility by rail make it necessary.

So how were the trees cut down in the first place?

Inspectors decide whether boards should be used for building or made into paper. ▶

Tree Time

T rees are made of cellulose fibers, which are long tubular connections of wood cells that often extend the length of the tree. Inside the tree, a natural hardening substance, lignin, makes the wood hard while sugars, resins, and oils help keep the tree healthy and growing. The only parts of a tree that are living and changing are the leaves, the upper parts of the shoot, the ends of the branches, roots, and the cambium, which is a thin layer of cells under the bark. When a tree grows, the most noticeable growth occurs at the very tips of the stems and branches and in its diameter. The diameter gets bigger when the cells in the cambium divide to create wood to the inside and bark to the outside of it.

People can determine the age of a tree by looking at a cross-section of the trunk. The cross-section shows growth rings. Each growth ring marks one year in the life of the tree.

RISING TO A CHALLENGE

How can huge trees be moved if trucks cannot get into an area?

When the tree growing area is steep, or difficult to reach by trucks, special equipment must be used. Sometimes hot-air balloons or helicopters are brought in to lift the heavy logs.

In other instances, a system called high-lead logging is employed. It uses pulleys and steel cables to drag logs up sharp inclines or swing them across ravines. Movable towers powered by electric, steam, or diesel winches support the pulleys and cables.

High-lead logging can transport logs to sites as far away as 1,500 feet. To drag the logs farther, systems using cables called skylines are used to raise entire logs off the ground and move them.

The development of these technologies has made the work of loggers much easier. In the past, loggers relied on the power of rivers and streams to transport logs. Logs were cut and stacked by the edge of the water until the spring thaw. The logs were then thrown into the water, followed by workers called lumberjacks, who wore spiked boots and held sharp harpoons to help drive the logs toward their destination. These log drives were very dangerous. They created enormous logjams in the water and often caused a lumberjack to lose a limb.

Today, in parts of the world where rails, roads, and technologies are not as readily available, water is still used to transport lumber in this way.

▲ *An early sheet of paper found in the ruins of a spur of the Great Wall of China and dated around A.D.150.*

Loading lifting logs ▼

Paper Providers

Ts'ai Lun, a Chinese diplomat and scholar, first made paper by grinding up plants, such as mulberry bark, linen, and hemp. Then he spread out the mixture in a coarse mat and bamboo frame and let it dry in the sun.

In the 1700s, Frenchman Rene de Reamur added another important piece to the paper-making process by watching insects called paper wasps. As he observed the wasps munch on wood and spit out the mush to make nests, Reamur had the notion that they were making paper from wood. He was the first to realize that wood could be broken down and then put back together as paper.

Trees Cut by Loggers in Forest

How were the trees cut down in the first place?

Trees are cut in a forest by powerful machines called tree shears and tree harvesters. Tree shears are bladed machines, attached to tractors, that work like enormous scissors to cut through tree trunks. Tree harvesters are so big that they can cut down a tree, remove the branches, cut it into logs, and even sort the logs.

Loggers, workers who cut down the trees, only chop trees in specific parts of a forest. To preserve areas for wildlife and recreational use, logging companies grow tree farms, called managed timberlands. In these areas, certain kinds of trees are chopped down while others are left standing. After the trees are cut, the loggers skid, or drag, the logs to a central place in the woods known as the landing. There, they cut the trees down further so trucks, trains, or boats can ship the wood to the sawmill.

But which trees make the best paper?

Loggers examining cut logs (top) and guiding them down river (bottom) to the saw mill ▶

▲ *Clear-cut area (background) and forested area (foreground)*

Cutting into Controversy

Clear-cutting trees is the practice of removing all the trees in a selected area. It began in the nineteenth century as an economical means of getting large amounts of usable wood to a sawmill. After the ruin of many forests and natural habitats, loggers began to use selective management to cut the most usable trees without totally ruining a forest. Today, clear-cutting is still a controversial subject. Some experts believe that clear-cutting creates the best conditions for regrowing trees because it provides complete sunlight for new growth. Other foresters, however, oppose this practice. They say it degrades the environment because it removes soil from the site, destroys wildlife habitats, increases the severity of flooding, and leaves behind an ugly, scarred landscape.

Specific Kinds of Trees Are Grown for Papermaking

Which trees make the best paper?

Certain trees in the forest are grown just for papermaking. Loggers know which trees these are, and where they are located.

Papermaking trees include birch, fir, gum, hemlock, oak, pine, and spruce. The trees are further classified as either hardwood or softwood. Hardwood trees, such as oak, have shorter fibers than softwoods. Though the hardwood birch tree is suitable for papermaking, most paper made from hardwoods would be exceptionally smooth, but too thin for some uses. Softwood trees, which are conifers such as pine and spruce, have longer wood fibers, which create stronger paper. The surface of such paper, however, would be too rough for printing or writing. Because of these characteristics, papermakers use a blend of hardwood and softwood fibers. Using different amounts of both fibers in a pulp mixture allows papermakers to produce papers with different combinations of strength, texture, and color.

In managed timberland, new trees are planted to replace trees that are cut. This practice is called sustainable forestry because it enables foresters to maintain a good supply of trees.

So how are new trees planted?

Hardwood trees such as birch (left) make smooth paper; softwoods such as pine (center) and spruce (right) make strong paper. Papermakers often use a blend of hardwoods and softwoods.

Map of U.S. Trees

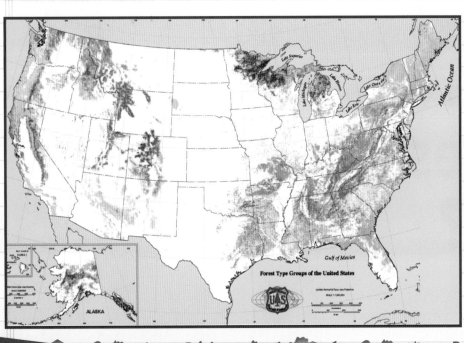

Forest Type Groups of the United States

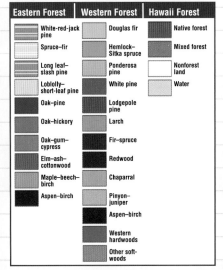

Eastern Forest	Western Forest	Hawaii Forest
White-red-jack pine	Douglas fir	Native forest
Spruce-fir	Hemlock-Sitka spruce	Mixed forest
Long leaf-slash pine	Ponderosa pine	Nonforest land
Loblolly-short-leaf pine	White pine	Water
Oak-pine	Lodgepole pine	
Oak-hickory	Larch	
Oak-gum-cypress	Fir-spruce	
Elm-ash-cottonwood	Redwood	
Maple-beech-birch	Chaparral	
Aspen-birch	Pinyon-juniper	
	Aspen-birch	
	Western hardwoods	
	Other softwoods	

New Trees Planted as Seeds and Seedlings

How are new trees planted?

Foresters plant new crops of trees as seeds and seedlings, or young trees. Seeds that have been treated with chemicals to discourage animals from eating them are usually planted in the late fall or early spring in the managed timberland forests. Airplanes or helicopters usually scatter hundreds or thousands of seeds at once.

Trees also grow from artificial reforestation. In this process, seedlings grow in a nursery—a separate plot of land where young trees develop. The land in the nursery has nutrient-rich soil, which is watered constantly to ensure plant growth. After a period of time ranging from one to four years, when the seedlings grow to a certain height and strength, they are transplanted into the forest. When these seedlings are planted in an area that has never held trees before, it is called afforestation.

Truck unloading young trees ▶

Consumers' Used Paper Products Are Sent to Recycling Plant to Be Sorted	Paper from Wood Used to Make Original Newspaper	Three Processing Methods Turn Wood to Pulp	Shaved, Cut Wood Travels to Paper Mill from Sawmill	Trees Cut by Loggers in Forest	Specific Kinds of Trees are Grown for Paper-Making	New Trees Planted as Seeds and Seedlings

Plant a Seedling

Trees provide shade and produce food and shelter for wildlife. They help to filter pollution from the air, prevent soil erosion, and add beauty to the land. These are just a few of the many reasons to plant a tree.

Generally, it is best to plant a tree in the fall before the ground freezes, or early in the spring before the tree shows signs of growth. To plant a tree, do the following:

- First, buy a seedling from a local nursery or a mail-order business. Be sure to choose a tree that is native to your area and will grow in the space you have available. Employees of the nursery can help you make a good choice.

- Choose a planting site with nutrient-rich soil that will soak up rain.

- Dig a hole big enough to cover the seedling's roots. The hole should be at least three times as wide as the root ball.

- Handle the seedling by the base of the trunk. Be careful not to bruise the bark. Place the root collar (the area where roots join the stem) slightly below ground level.

- The roots of a tree are often tied or wrapped. Remove wrapping before filling the hole with dirt.

- Settle the tree with water to avoid the formation of air pockets. Then cover the roots with soil and tap firmly.

- Add 2-3 inches of mulch—finely ground organic matter, such as wood and plants—to the area around the tree, but stay clear of the trunk by about 6 inches.

- After mulching, the tree needs to be watered regularly to ensure that the tree takes hold.

- For the next few years, the seedling will grow best with care. Water and weed the area around it, and periodically add mulch to the soil.

The Big Picture

Explore the science behind recycled newspapers, step by step:

13 Readers Open Newspapers Sold at Newsstands

People read newspapers, often made of recycled newspaper.
(pages 6-7)

10 Pulp Dries into Sheets of Paper

At the paper mill, metal rollers squeeze moisture from the wet pulp. Hot rollers then dry it completely.
(pages 14-15)

8 Recycled Paper Comes from the Recycling Plant

The recycling plant delivers old paper in huge bales to the paper mill.
(pages 18-19)

12 Trucks Deliver Newspapers to Sellers and Distributors

Trucks pick up papers from the printing plant and deliver them to various locations..
(pages 8-9)

11 Newsprint Comes from a Paper Mill

Huge rolls of newsprint come to the printing plant from the paper mill.
(pages 12-13)

9 Pulp Mixes with Water and Chemicals for Cleaning and Bonding

Scraps of old or used paper are placed in a pulping machine to make pulp. The pulp is cleaned and de-inked.
(pages 16-17)

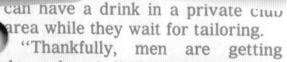

7 Consumers' Used Paper Products are Sent to Recycling Plant to Be Sorted

Recyclers get used paper products from consumers and sort it.
(pages 20-21)

5 Three Processing Methods Turn Wood to Pulp

At a paper mill, three process styles may be used to convert wood into pulp: mechanical, chemical, and semi-chemical.
(pages 28-29)

3 Trees Cut by Loggers in Forest

Loggers use machines called tree shears and tree harvesters to cut trees.
(pages 34-35)

6 Paper from Wood Used to Make Original Newspaper

New paper mills make original paper from wood.
(pages 24-25)

4 Shaved, Cut Wood Travels to Paper Mill from Saw Mill

Sawmill machines shave and cut wood into lumber, which is shipped by rail, boat, or truck to the paper mill.
(pages 30-31)

2 Specific Kinds of Trees Are Grown for Papermaking

Trees for papermaking, including birch, oak, pine, and spruce, come from managed timberlands. Papermakers know to combine hardwood and softwood.
(pages 36-37)

1 New Trees Planted as Seeds and Seedlings

Foresters produce new trees in a forest by planting either seeds or seedlings grown in a nursery.
(pages 38-39)

Tree to Paper Timeline

Papermaking has a long history

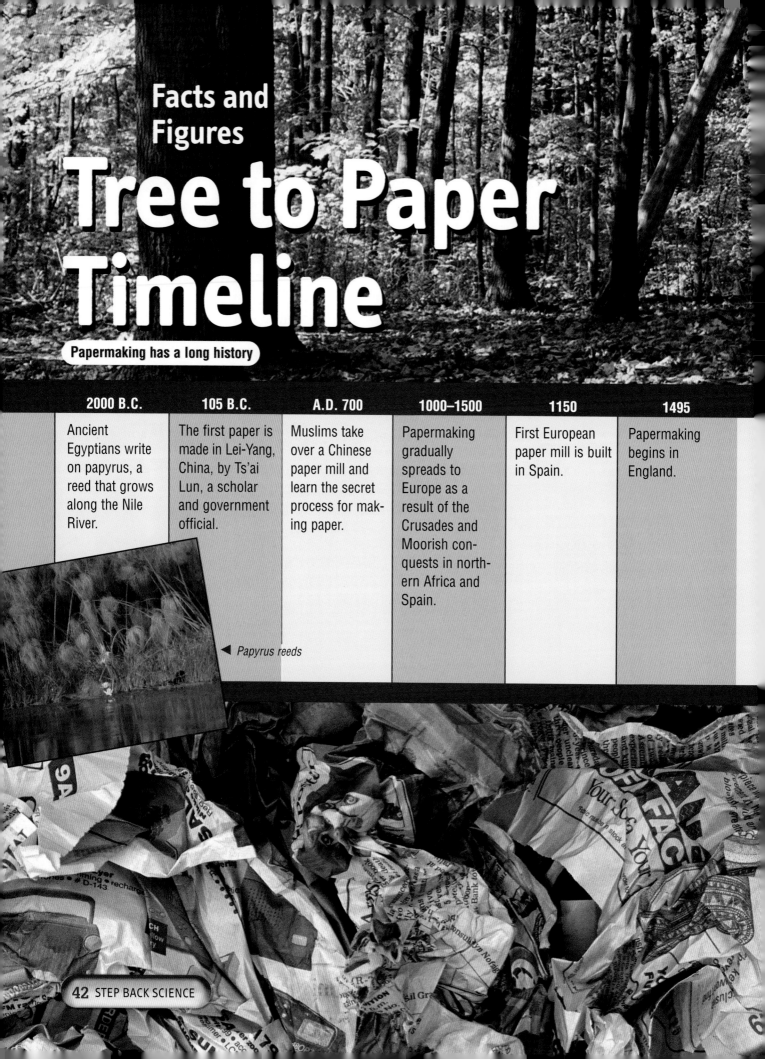

2000 B.C.	105 B.C.	A.D. 700	1000–1500	1150	1495
Ancient Egyptians write on papyrus, a reed that grows along the Nile River.	The first paper is made in Lei-Yang, China, by Ts'ai Lun, a scholar and government official.	Muslims take over a Chinese paper mill and learn the secret process for making paper.	Papermaking gradually spreads to Europe as a result of the Crusades and Moorish conquests in northern Africa and Spain.	First European paper mill is built in Spain.	Papermaking begins in England.

◄ *Papyrus reeds*

1690	1798	1840	1866	1911
First American paper mill is built in Germantown, Pennsylvania.	Frenchman Nicholas Louis Robert invents a machine to make paper in continuous rolls instead of sheets. The Fourdrinier brothers, English merchants, finance improvements to this machine.	Stone groundwood method for making wood pulp is developed in Germany.	American Benjamin Tilghman develops a process for chemically produced pulp.	Kraft process becomes the dominant pulping process.

▲ Paper mill, Germantown, Pennsylvania

◄ Fourdinier rolls on paper machine.

Fast Paper Facts

- In 1999, U.S. paper mills produced 11.85 million tons of newsprint.
- At least 70 percent of all newsprint used in the United States is recovered for recycling.
- Recovered fiber makes up one-third of the total fiber used to make the world's paper.
- In 2000, U.S. recycling plants recovered nearly 48 percent of all paper used. More paper is now recovered in the United States than is sent to landfills.
- Approximately 86 percent of all Americans have access to some form of paper recycling program.

Wonders and Words

Q: *Are most trees cut down to make paper?*

A: Only about 28 percent of the trees cut down in a year are used for paper. Forty-nine percent are used for building and construction purposes, while 23 percent are used for heating and cleaning agents (from resins or other liquid substances made from trees).

Q: *How much of a tree is used when it is cut down?*

A: Nearly all of it. The trunk is used for boards and planks. Often, scraps from cutting and processing lumber can be used for papermaking. Bark is used for fuel, and wood chemicals are extracted for making products such as pine solvents, cleaning agents, turpentine, and gums.

Q: *Why do people recycle paper?*

A: Not just to save trees, but also to save landfill space. Even with the existence of successful recycling programs, paper still makes up about one-third of all the materials that go into a landfill—a plot of land on which garbage is dumped. Landfills are expensive to maintain and take up valuable space. By recycling paper, people slow the rate at which landfills reach their capacity. The process of recycling paper also produces fiber, which can be used for other new paper or cardboard.

Glossary

Afforestation: Process of planting seedlings in an area that has never been covered by trees before

Artificial Reforestation: Process of creating a new forest by planting seeds or seedlings

De-inking: Process of removing inks and adhesives from recovered paper

Grindstone: Flat, circular stone that revolves on an axle and is used for grinding or shaping

Kraft process: Chemical process of converting wood to pulp by cooking wood chips in a solution of caustic soda and sodium sulfide

Lignin: Substance in cell walls of wood that gives wood its hardness

Managed timberlands: Tree farms that allow foresters to maintain an adequate supply of trees

Pre-consumer recovered paper: Trimmings and scraps from printing, carton manufacturing, and other converting processes that have never been handled by consumers

Post-consumer recovered paper: Paper that has been used by consumers such as newspapers, magazines, and used cartons

Pulp: Macerated wood, or fibrous material, prepared from wood or recovered paper for use in manufacturing paper

Seedling: Young or immature tree grown from seed; a tree or nursery plant not yet transplanted into the ground

Stickies: Sticky substances such as glue that are removed from recovered paper.

Stone groundwood process: Mechanical process of pulping wood developed in 1840 in which logs are thrust against a grindstone

Sulfite process: Chemical process of converting wood chips to pulp by cooking them in an acid solution.

Sustainable forestry: Practice of planting new trees to take the place of trees that have been cut down

Index

Credits:

Produced by: J. A. Ball Associates, Inc.
Jacqueline Ball, Justine Ciovacco
Daniel H. Franck, Ph.D., Science Consultant

Art Direction, Design, and Production:
designlabnyc
Todd Cooper, Sonia Gauba

Writer: Susan Korman

Answer to question on page 19:
c. 5-7

Cover: Brooke Fasani: boy reading newspaper; Photo Researchers, Inc.: paper on rollers; PhotoDisc, Inc.: rolls of paper, logs; ArtToday: seedling.

Interior: Photospin: p.3, pp.42-43 crumpled newspaper, p.13 scissors, p.29 papyrus sheet, pp.40-41 newspaper (background), pp.42-43 forest, p.42 papyrus plants, p.44 wood (background); Brooke Fasani: pp.6-7 boy reading newspaper, p.9 stack of newspapers, p.19 girl with tree, p.21 boy stacking newspapers, empty boxes, p.44 stacks for recycling; Ablestock/Hemera: p.8 eggs, p.15 crayons, p.17 cardboard boxes, p.23 recycling bins, forest, p.26 forest, pp.30-31 wood mill, pp.32-33 forklift loading logs, pp.36-37 birch tree, pine tree, spruce tree, pp.38-39 truck with seedlings; PhotoDisc, Inc.: pp.10-11 meeting, pp.12-13 rolls of paper, pp.18-19 bales of paper, p.25, 27 wood chips, p.26 logs, pp.34-35, logs on water, loggers in boat, p.35 clear-cut forest; Library of Congress: p.11 early printing press, p.43 papermill, Fourdinier; Photo Researchers, Inc.: pp.14-15 paper on rollers; Blackbirch Press Archives: p.17 de-inking process; Visuals Unlimited, Inc.: p.27, 29 pulp; ArtToday: p.29 papyrus reeds, p.31 cross-section of tree, p.33 early paper sample, p.37 forest density map, p.39 seedling, p.43 pulping

For More Information

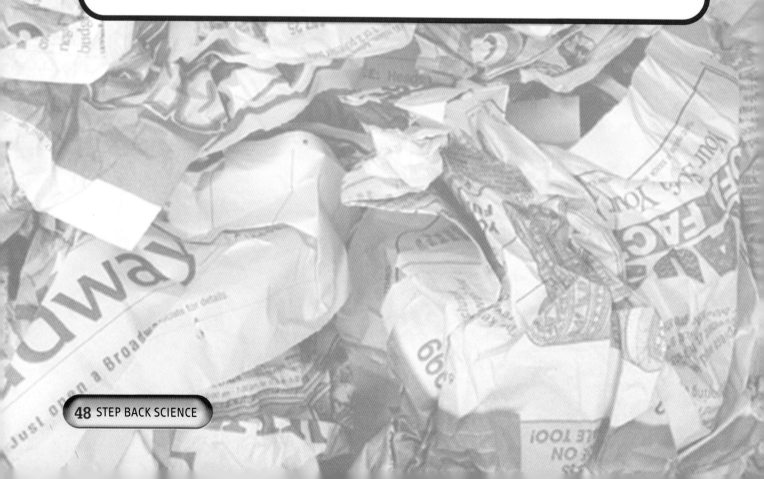

Sponsored by the American Forest and Paper Association

www.afandpa.org

Offers articles, statistics, facts, and figures about the U.S. lumber and paper industries, as well as downloadable activities for students and teachers.

Sponsored by the American Forests Organization

www.americanforests.org

Promotes planting trees all over the world through a wide variety of activities, programs, and projects.

Condon, Judith. *Recycling Paper.* New York: Franklin Watts, Inc., 1990.

Jonas, Gerald. *The Living Earth Book of North American Trees.* Pleasantville: Readers Digest Books, 1993.

Woods, Samuel G. *Recycled Paper from Start to Finish.* Woodbridge, CT: Blackbirch Press, 2000.